D1411889

Understanding the Elements of the Periodic Table™

GOLD

Brian Belval

rosen central™

The Rosen Publishing Group, Inc., New York

For Mom and Dad

Published in 2007 by The Rosen Publishing Group, Inc.
29 East 21st Street, New York, NY 10010

First Edition

Library of Congress Cataloging-in-Publication Data

Belval, Brian.
Gold / Brian Belval.—1st ed.
p. cm.—(Understanding the elements of the periodic table)
Includes bibliographical references and index.
ISBN 1-4042-0708-2 (library binding)
1. Gold—Juvenile literature. 2. Periodic law—Tables—Juvenile literature.
I. Title. II. Series.
QD181.A9B45 2007
546'.656—dc22

2005028664

Manufactured in the United States of America

On the cover: Gold's square on the periodic table of elements; the atomic structure of a gold atom *(inset)*.

Contents

Introduction

The elements are the building blocks of everything in the universe. They make up the planets of our solar system, the air we breathe, the ground we walk on, our bodies, and the shoes on our feet.

There are more than 100 different elements. A quick field trip to the beach will introduce us to many of them. This is a mental field trip, so a permission slip from a parent or guardian is not required.

Imagine a perfectly sunny day and a cool breeze coming off the water. Take off your shoes and feel the sand beneath your feet. Did you know the sand is made of the elements silicon (Si) and oxygen (O)? The air you are breathing is mostly oxygen and nitrogen (N), with small amounts of argon (Ar), krypton (Kr), hydrogen (H), carbon (C), helium (He), xenon (Xe), and radon (Rn). Let's contemplate that while we walk to the concession stand. Lucky for you, it is stocked with an assortment of elements. The salt on your french fries consists of the elements sodium (Na) and chlorine (Cl). Hot dogs are delicious, but it will take an entire paragraph to list all the elements in a hot dog. They include nitrogen, carbon, hydrogen, oxygen, potassium (K), phosphorus (P), and iron (Fe), to name several.

And then there is gold. You can find it in the earrings, necklaces, and other pieces of jewelry worn by the beachgoers. You can even find it in some people's mouths. That's right—gold is used to repair and replace damaged teeth.

Nuggets of gold, such as the one above, are extremely rare. More often, gold is found in nature in small grains or flakes. The world's largest gold nugget was discovered in 1869 in Australia. Known as the Welcome Stranger nugget, it weighed 159 pounds (72 kilograms) and measured nearly 2 feet (60 centimeters) in length.

There are other surprising places where you can find gold at the beach. For example, many cell phones have small amounts of gold inside them to protect electronic connections. There is also gold in the ocean. Most people don't know it, but the world's oceans contain more than 10,000,000 tons (9,000,000 metric tons) of gold. Unfortunately, it is present in very low concentrations: for every 100 billion parts of ocean, there is only one part gold.

Our field trip has come to an end, but now you can see that the elements are everywhere. Even gold, one of the least common elements, can be found in many different places. In the rest of this book, we will continue our search for gold. When we find it, we will look at it closely, like a scientist does. We will break it apart and study it from the inside out. The goal is to get to know gold and its fascinating chemistry.

Chapter One
The Atom

Gold has been known to and used by humans since ancient times. Prehistoric humans probably first discovered the bright flakes and nuggets of gold in a stream or riverbed. Early civilizations collected gold and learned how to work with it. Modern archaeologists have found gold jewelry made by skilled craftsmen more than 5,000 years ago in Egypt. The first gold coins were made about 2,500 years ago in Asia Minor, an area that today is the country of Turkey.

For much of history, gold was known to be a metal but was not considered an element. Early chemists, known as alchemists, believed that gold and other metals were mixtures of four fundamental elements: earth, air, water, and fire. By altering the proportions of these four elements, they believed they could change one metal into another. The key to this transformation was a magical substance called the philosopher's stone. However, there is no such thing as the philosopher's stone. The alchemists' quest to create gold out of less valuable metals was bound to fail.

By the eighteenth century, the four-element theory had been discarded. Scientists began defining an element as any substance that could not be broken down into simpler substances. Water was not an element because it could be broken down into two other substances: hydrogen and oxygen. Other suspected elements were run through many tests. They were heated and cooled, doused with acid, pounded, and filtered. By

Many archaeologists and historians believe that the world's first coins were minted in the kingdom of Lydia, in Asia Minor (modern Turkey). Pictured above are the two sides of a coin called the Lydian third stater. It dates from about 600 BC. The coin is made out of electrum, a special metal made up of both silver and gold.

the beginning of the nineteenth century, samples were also being jolted with electricity. If the sample broke down into something else, it was not an element. Gold, however, was an element. So were copper (Cu), lead (Pb), tin (Sn), hydrogen, oxygen, and many others.

Inside the Elements

Scientists began to suspect that the elements were made of tiny individual particles, which they called atoms. They performed experiments to test this theory. The bulk of these experiments involved carefully weighing the amount of each element that combined to form a compound. In one famous experiment, John Dalton (1766–1844), a chemist from England, weighed the amounts of carbon and oxygen that reacted to form two different compounds. He found that twice as much oxygen combined with the same amount of carbon in one compound compared to the other. This result could be explained if one of the compounds was made out of twice as many atoms of oxygen as the other compound. Additional tests supported this theory. These experiments provided solid evidence that elements were made out of a basic unit, or building block, known as the atom.

However, there was still one unanswered question. If the elements were made out of atoms, then what made up the atoms? By the early twentieth century, scientists had found the answer to this question. They discovered that atoms are made of three smaller particles: electrons, protons, and neutrons.

Protons

Protons are positively charged. Together with neutrons they form the center of the atom, which is known as the nucleus. The nucleus is incredibly small but also incredibly dense. According to Steven Zumdahl, a professor of chemistry at the University of Illinois, if a gold

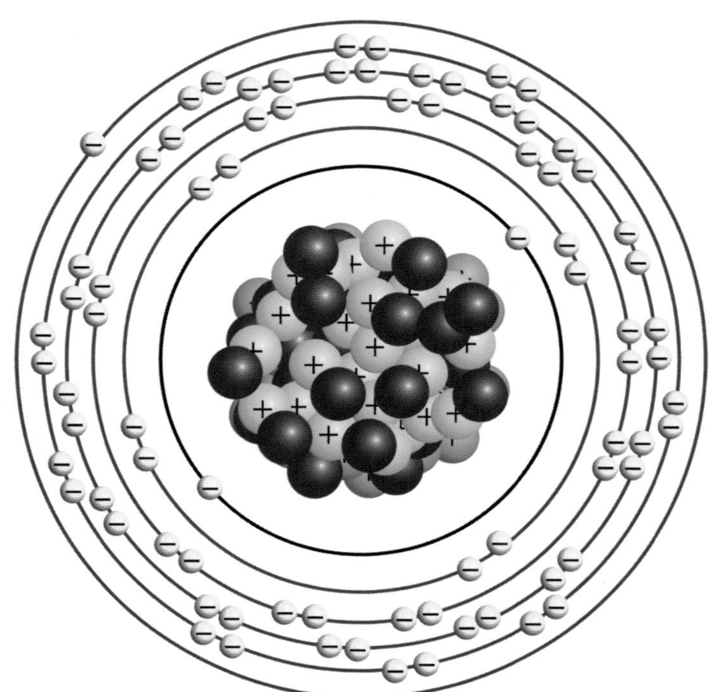

Protons (green) and neutrons (black) make up the nuclei of all atoms. Electrons (light blue) travel in shells outside the nucleus. A gold atom has six shells of electrons, as can be seen in the diagram at left.

nucleus were the size of a pea, it would weigh 250,000,000 tons (227,000,000 metric tons)!

The number of protons in an atom determines the element. For example, an atom of gold always has 79 protons. Take away one proton and you don't have gold anymore—you have platinum (Pt). Add one proton to the atom and you have the element mercury (Hg).

Neutrons

Neutrons have neither a positive nor a negative charge. They are neutral. They act as the glue of the nucleus, keeping all the particles tightly bound. Without the neutrons, the protons would not get close to one another. This is because protons are positively charged, and particles with the same charge repel each other, or push each other away. The neutrons work by stepping between the protons and neutralizing their repelling forces.

Electrons

Electrons are negatively charged particles. They are attracted to the positively charged protons in the nucleus. The electrons dart around outside the nucleus in what are known as orbitals, or shells. Electrons are much lighter than protons and neutrons. They contribute very little to the mass of an atom.

An atom can have anywhere from one to seven shells of electrons. Each shell is progressively farther away from the nucleus. Electrons in the inner shells are held very tightly by the attractive forces of the protons in the nucleus. The electrons in the outer shells are held less tightly.

An Atom of Gold

Each atom of gold has 79 protons, 79 electrons, and 118 neutrons. The number of protons equals the number of electrons in order to balance out

Gold Snapshot

79 197
Au

Chemical Symbol:	Au
Classification:	Transition metal
Discovered By:	Element has been known since prehistoric times
Atomic Number:	79
Atomic Weight:	196.966 atomic mass units (amu)
Protons:	79
Electrons:	79
Neutrons:	118
Density at 68°F (20°C):	19.3 g/cm^3
Melting Point:	1,948°F (1,064°C)
Boiling Point:	5,173°F (2,856°C)
Other Properties:	Conducts electricity, malleable, ductile
Commonly Found:	In streambeds or riverbeds, veins below Earth's surface

their charges. When the protons and electrons are not balanced, the atom is said to be an ion. An ion is an atom, or cluster of atoms, that has either gained or lost electrons. An ion has a positive charge if it has lost one or more electrons. It has a negative charge if it has gained one or more electrons. For example, a gold atom that has lost three electrons is an ion with a charge of +3.

Gold has six shells of electrons. The first shell has two electrons, while there are eight in the second, eighteen in the third, thirty-two in the fourth, and eighteen in the fifth. The sixth and outer shell has only one electron.

Chemical Bonding

The outer-shell electrons, also known as valence electrons, are of special interest to chemists. These electrons are involved in chemical bonding, which occurs when two or more atoms join together. When atoms of different elements bond, they form a compound. Water, salt, and sugar are a few examples of compounds. Gold compounds will be discussed in chapter 5.

As mentioned, gold has one outer-shell electron. Often, elements with one electron in their outer shell are very reactive and easily combine with other elements to form compounds. This is true of such elements as sodium, potassium, and lithium (Li). However, gold's relatively large nucleus exerts a strong pull on the lone electron in the outer shell. This means that the electron is pulled in very tightly and is usually not made available for bonding.

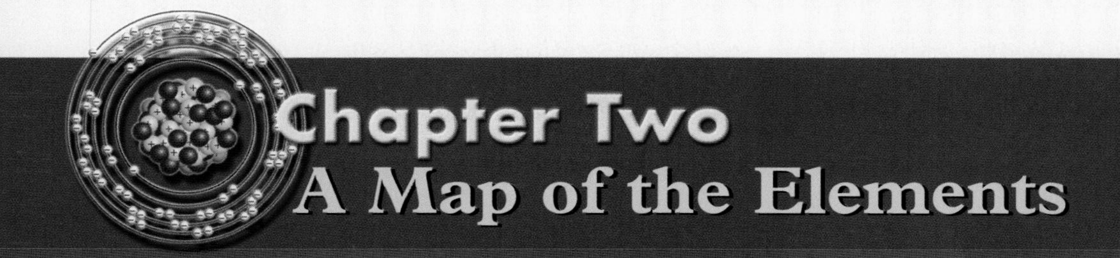

Chapter Two
A Map of the Elements

In 1869, Dmitry Mendeleyev (1834–1907), a Russian chemistry professor, created the first periodic table of elements. His table arranged the known elements by weight, from lightest to heaviest. He included it in a chemistry textbook to help his students recognize patterns in the properties of the elements.

Although today's periodic table has been revised and expanded since the days of Mendeleyev, the basic look is still the same. (See the periodic table on pages 42–43.) Each element is given a one-, two-, or three-letter symbol. Usually, the symbol is an abbreviation of the element's name in English. H, for example, stands for hydrogen. Gold is a little bit of an odd-ball, though. Its symbol is Au, short for *aurum*, which is what gold is called in Latin. The symbols of a few other elements are based on their Latin names, also. For example, the symbols for iron (Fe) and sodium (Na) come from the Latin words *ferrum* and *natrium*.

The elements in the table are arranged in rows, known as periods, and columns, known as groups. The periods are numbered one to seven from top to bottom. The groups are numbered in one of two ways, depending on which version of the table you are using. In the first numbering system, the groups are numbered from one to eighteen going from left to right. In the second system, each group is given a Roman numeral followed by a letter. For example, gold is in group IB, while sulfur (S) is in group VIA.

Dmitry Mendeleyev *(right)* published a classification of the sixty-two chemical elements known during his time. He grouped them into families, leaving gaps where he believed elements should belong once they were discovered.

Below the main table is a smaller table made up of two rows and fourteen columns. This block is actually part of the main table. The first row of the smaller table should go on the main table in between lanthanum (La, atomic number 57) and hafnium (Hf, atomic number 72). The second row of the smaller table goes between actinium (Ac, atomic number 89) and rutherfordium (Rf, atomic number 104). However, when the smaller table is included within the larger table, the resulting chart is inconveniently wide.

The Groups

Elements within a group often have similar chemical properties. This means that they react with other elements or compounds in similar ways. For example, the elements in the first two columns react with elements in column VIIA to form a type of compound called a salt. Table salt, or sodium chloride (NaCl), is the best known of the salts.

Some of the groups of the table have been given special names. The elements in the first column are known as the alkali metals, and the elements in the second column are known as the alkaline earth metals. The elements in column VIIA are known as the halogens. The far right column

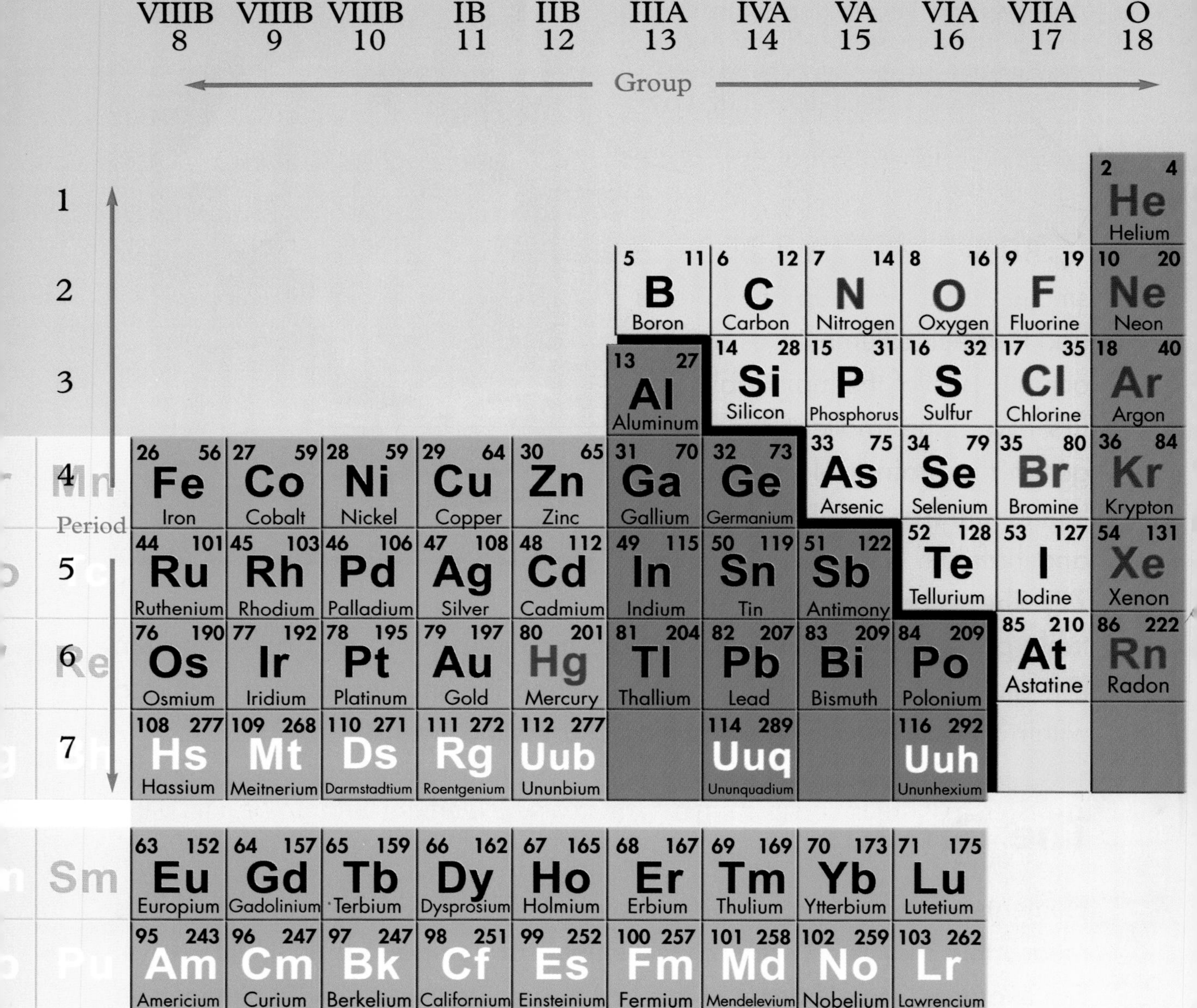

The periodic table consists of seven periods (rows) and eighteen groups (columns). This portion of the periodic table shows groups 8 through 18. The line shown separating the blue and yellow elements is known as the staircase line. Most of the elements to the left of the staircase line, including gold in group 11, are metals. Most of the elements to the right of the staircase line are classified as nonmetals.

of the table is made up of the noble gases. The elements in this group are unique because they never react with other elements.

Gold belongs to a block of elements called the transition metals. These elements form the middle of the table, from the third column to the twelfth column. The transition metals do not share chemical properties as neatly as other elements that are grouped together. They do, however, have some things in common. The transition metals are able to conduct electricity and heat. They also have the ability to be twisted into wires and pounded into sheets. In addition to gold, other well-known transition metals include iron, nickel (Ni), copper, zinc (Zn), and silver (Ag).

The Numbers on the Table

If you look at any square on the periodic table, you will see two numbers. The smaller number is the atomic number, which is equal to the number of protons in an atom of the element. For gold, the atomic number is 79, which means an atom of gold has 79 protons. Copper has an atomic number of 29, so an atom of copper has 29 protons. Unless an atom is an ion, its atomic number also indicates the number of electrons.

If you scan the table, you will notice that the elements are arranged from left to right by their atomic numbers. The element with an atomic number of 1 (hydrogen) is in the upper left-hand square. Ununhexium, the element with the largest atomic number, is in the bottom right-hand square.

The larger number in each square is the atomic weight (also known as atomic mass). The atomic weight is the average weight of an atom of the element, measured in atomic mass units (amu). Gold has an atomic weight of 196.966 amu. For convenience, this number is often rounded to 197.

Atomic weight is an average measurement because elements are not always made out of identical atoms. Often, elements consist of isotopes, which are atoms with the same number of protons but different numbers of neutrons. Carbon, for example, has three naturally occurring isotopes.

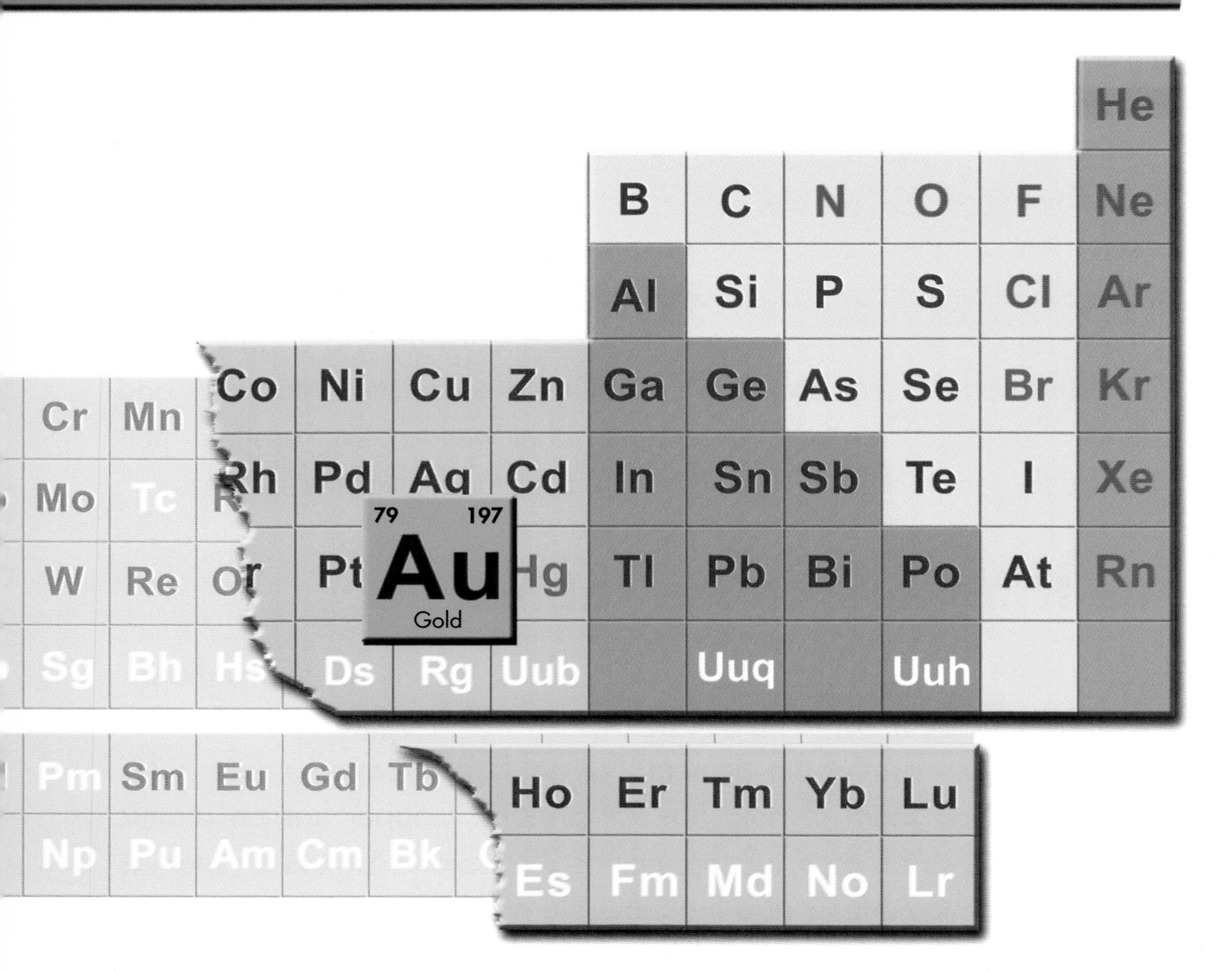

With an atomic weight of approximately 197 atomic mass units, gold is one of the heaviest naturally occurring elements. Many of the elements heavier than gold, such as einsteinium (Es), fermium (Fm), and lawrencium (Lr), are synthetic. They do not occur in nature and instead are made by scientists in a laboratory.

The most common isotope of carbon has twelve neutrons, while another has thirteen neutrons. A third isotope has fourteen neutrons. The atomic weight is the average weight of all these isotopes, taking into consideration how often each isotope occurs. Gold is somewhat unusual because it has only one naturally occurring form, which always has 118 neutrons.

Work It Out

Here's a quick practice exercise. You'll need to refer to the periodic table on pages 42–43 to figure this one out.

How many protons and electrons are there in an atom of selenium (Se)?

The number of protons is equal to the element's atomic number. Selenium's atomic number is 34, so it has 34 protons. Each atom of selenium also has 34 electrons to balance out the positive charge of the protons.

What Can the Periodic Table Do for You?

In summary, the periodic table is useful because it is a compact and convenient chart providing a great deal of information about each of the elements. The table allows you to quickly find out how many protons and electrons are in an atom of gold—or any of the elements. You can also determine the number of electron shells that an element has, since this number is equal to the period number.

Because the table arranges the elements into groups, you can make judgments about an element simply by its location in the table. In the case of gold, which is located with other transition metals, you can assume that gold will have some properties shared by those elements. Properties of transition metals include the ability to conduct electricity and heat, a shiny surface, and the ability to be pulled into wires and pounded into sheets.

As you can see, the periodic table is an important resource. You might even call it the map of the elements. Without it, any aspiring chemist is sure to get lost.

Chapter Three
The Properties of Gold

Imagine that you are a professional chemist. Your boss gives you an envelope containing a small metal cube that is believed to be gold. Your job is to test it. Is it really gold?

To go about this task, you would investigate the properties of the sample. In science, "property" is another way of saying "characteristic," or "trait." It is something that distinguishes one thing from another. Properties can either be measured or observed. They include qualities such as density, melting point, and color. By measuring and observing the properties of your sample, you would be able to determine if it was gold or something else.

Take a Look

The first thing you can do to identify your sample is to record what it looks like. Properties you should note include color, texture, and whether it is a solid, liquid, or gas. If your sample is gold, it should be a deep yellow color, have a shiny surface, and be a solid at room temperature. If it doesn't have these properties, you can be pretty sure it is not gold.

However, just because it looks like gold does not mean you are finished with your examination. Have you ever seen a fake gold watch or

Pictured above is a nugget of pyrite, a common mineral made of iron and sulfur. It is known as fool's gold because its brass-yellow color and shiny appearance have fooled many into believing it is real gold. Along with quartz, pyrite is often found near gold veins.

necklace? Just like on the street, looks can be deceiving inside the laboratory. There are many tests you can perform to confirm whether what you have is actually gold.

What It Means to Be a Metal

Gold, like all metals, can conduct electricity and heat, can be pounded into sheets, and can be stretched into wires. Your next tests should confirm whether your sample has these properties.

Electrical and Thermal Conductivity

Electricity is the flow of electrons through a substance. Gold can conduct electricity because its outer-shell electrons can easily move from atom to atom. When electricity is applied, the electrons begin to flow, creating an electrical current. The electrical conductivity of gold can be measured in the laboratory using a battery and an ammeter, a device that measures electricity.

In nearly the same way that gold conducts electricity, it will also conduct heat. This property is known as thermal conductivity. One simple way to test for thermal conductivity is to measure how fast an object heats up once it comes into contact with another hot object or hot liquid.

Imagine that you have a cup made of gold and a cup made of glass. If you were to pour boiling water into each, the gold cup would become hot more quickly than the glass cup. This is because metals, such as gold, conduct heat much better than glass does.

Malleability

The property of malleability means that a substance can be pounded into sheets or bent into different shapes. A diamond (pure carbon) is not malleable. When it is struck, it will not flatten, nor can it be bent with any instrument. However, most metals, including gold, are malleable. This property explains why an element like aluminum can be made into foil and why silver can be shaped into a medallion. Gold is the most malleable of all the elements. It can be pounded into sheets 1,000 times thinner than a human hair.

In the photo on the facing page, an artisan applies gold leaf to the dome of the West Virginia state capitol. Gold leaf is thinner than tissue paper and must be handled with extreme care. Coating objects with sheets of gold leaf is known as gilding. It creates the illusion that an object, such as a dome or a statue, is made out of solid gold.

Take a hammer to your sample and see how malleable it is. If it can be pounded into a sheet, then it might be gold. But if it shatters into a million pieces, you can report to your boss that the sample is definitely not gold. You can also ask your boss for a broom to clean up the mess you made.

Ductility

Ductility is the ability to be stretched into wires. Copper is an example of a ductile element. This property, along with its ability to conduct electricity, makes copper a popular choice for electrical wiring. Compared to copper, gold is even more ductile. In fact, gold is the most ductile of all the elements. In highly controlled experiments, researchers have stretched gold into nanowires, extremely small wires that can measure just one atom thick! Gold makes excellent electrical wire, but since it is so expensive, it is not widely used for this purpose.

If you had wire-making equipment in your laboratory, you could test your sample and see how well it can be stretched into a wire.

Properties of Different Metals

	Gold	Copper	Aluminum	Zinc
Conducts electricity	yes	yes	yes	yes
Density (g/cm^3)	19.3	8.9	2.7	7.1
Melting point	1,948°F (1,064°C)	1,981°F (1,083°C)	1,221°F (660°C)	786°F (419°C)

Reaching the Melting Point

Melting point is often used to help distinguish between two substances. The melting point is the temperature at which a substance changes from the solid phase to the liquid phase. Melting points can vary radically from element to element and between different substances. The melting points of metallic elements range from –36 degrees Fahrenheit (–38 degrees Celsius) for mercury to 6,170°F (3,410°C) for tungsten (W). Gold falls somewhere in between, with a melting point of 1,948°F (1,064°C).

A temperature of nearly 2,000°F (1,093°C) is much hotter than what you can reach in your oven at home. But imagine that your high-tech

Because it is rare and valuable, gold is often recycled. The recycling process begins with the collection of discarded materials that contain gold, such as computer circuit boards. The gold is extracted from the materials and purified. In the photograph above, a technician pours molten (melted) recycled gold into water to produce grains of gold.

laboratory has a furnace that can reach that temperature. (These types of furnaces really do exist.) Place your sample in the furnace and turn up the heat. If it melts at any temperature other than 1,948°F (1,064°C), give or take a few degrees, then your sample is not gold.

The Density of Gold

Gold is a dense element. Density is the measure of mass per volume. So something that is dense has a lot of mass packed into a small space. Density is commonly measured in grams per cubic centimeter (g/cm^3). Gold's high density is due to two things: the mass of each individual atom and the way the atoms are packed together.

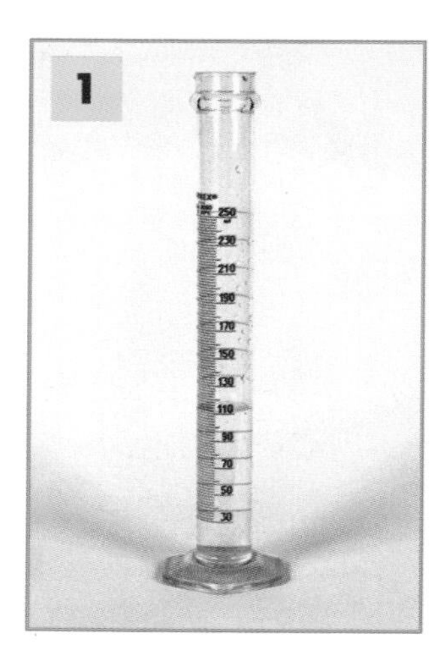

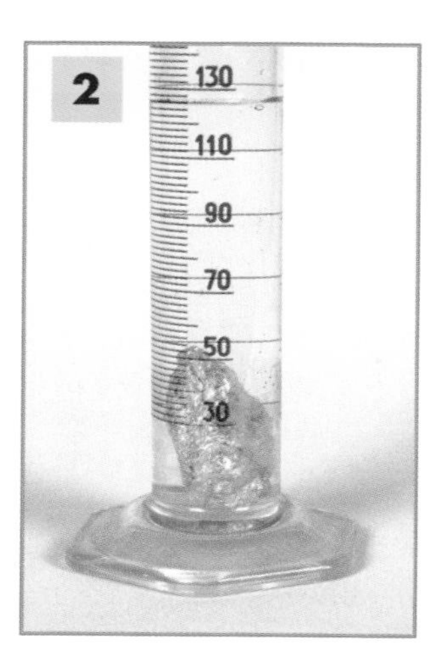

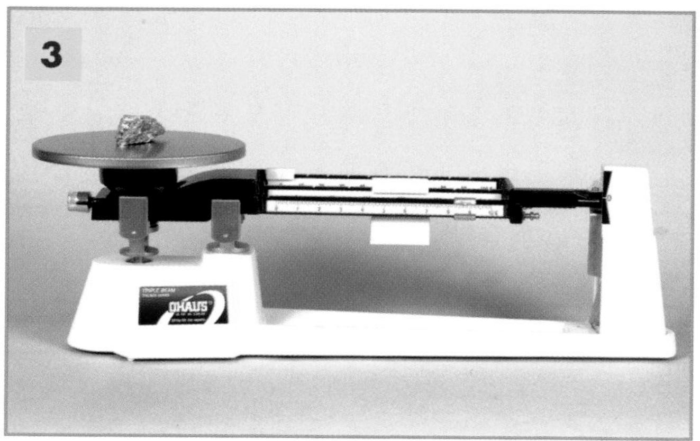

A graduated cylinder and a scale can be used to measure the density of gold. First, water is added to a graduated cylinder, and its volume is recorded (1). Then the gold sample is added to the cylinder, and the resulting volume is noted (2). The number recorded in (1) subtracted from the number recorded in (2) equals the volume of the gold. Finally, a scale is used to measure the mass of the gold sample (3). The density of the sample equals its mass divided by its volume.

First, let's look at a single atom of gold. Gold's nucleus of 79 protons and 118 neutrons has a strong effect on the surrounding electrons, pulling them in very tightly. The result is a tightly packed and very heavy atom.

But your sample of gold is more than just one atom—it is a collection of many atoms. In fact, a piece of gold the size of a grain of rice contains billions of atoms. Gold's density is also a result of all these atoms arranging themselves in tightly packed structures. In less dense elements, the atoms have more space between them. Each atom of gold, however, is crammed up against its neighbor, kind of like gumballs in a vending machine.

Measuring density in the laboratory is fairly straightforward. First, you need to weigh the sample using an accurate scale. Then you measure the sample's volume. Dividing the mass by the volume gives you the sample's density. If the sample is gold, its density should be about 19.3 g/cm^3.

Chemical Reactions

Another way to test an element is to see how it reacts with other elements or compounds. Fortunately for you, gold is very unique in this regard. Gold does not react with water, air, most acids, or almost anything else. In fact, gold is one of the least reactive of all the elements.

As mentioned earlier, gold tends not to be reactive because of the tight grip each nucleus has on its outer electron. When atoms of other elements pass by looking for electrons to grab and bond with, they are turned away. Atoms of gold tend to bond with each other rather than with atoms of any of the other elements.

So, one way to test your sample is to see what happens when you combine it with other substances. Try mixing your sample with water, acid, alcohol, laundry detergent, salt, soda pop, or anything you want. If it is gold, then none of these substances will change your sample.

Unlike many other metals, gold does not react with common acids. In the first beaker, a sample of zinc had been added to hydrochloric acid. The zinc reacts with the acid, producing bubbles of hydrogen gas. In the second beaker, gold does not react with hydrochloric acid and remains unchanged.

Name That Element

OK, so you have completed all the tests on your sample. It is a deep yellow solid with a shiny surface. It is able to conduct electricity. It can be pounded into a thin piece of foil and stretched into a wire. It melts at about 1,948°F (1,064°C) and has a density of 19.3 g/cm^3. If all these things are true, you can be certain the sample is gold.

Chapter Four Discovering Gold

Since the dawn of history, gold has been one of the elements most highly valued by humans. A major reason why gold is so valuable is its rarity. If you were to rank the elements from most abundant to least abundant, gold would be 73rd out of the 111 elements listed in the most widely accepted periodic table.

But it is also the physical and chemical properties of gold that make it so coveted. Its color and reflective surface make it visually appealing. Since it is malleable and ductile, artists and craftsmen can take a nugget of gold and shape it into something even more beautiful. And because gold does not react with air, water, and most other substances, it does not corrode, rust, or otherwise change over time.

Gold's value has led many to go to great lengths to retrieve it from the earth. It is not easy to find, so people looking for gold need special skills, equipment, and a lot of patience and perseverance.

Where Is the Gold?

People who search for gold are known as prospectors. They look for gold in two places—within Earth's crust and at the bottom of streams and rivers. Sections of Earth's crust with high concentrations of gold are known as veins. A vein with an exceptionally high concentration of gold is called

The Not-So-Pretty History of Gold

For thousands of years, gold has been used to make such beautiful objects as statues and jewelry. These golden treasures celebrate life and human achievement. However, there is a darker side to gold. Throughout history, the desire for gold has also brought out the worst in human behavior.

Humans' quest for gold has resulted in wars, death, destruction, and despicable acts of violence. Perhaps the best-known examples come from the history of Spanish conquistadores in the Americas during the sixteenth century. The conquistadores (Spanish for "conquerors") were soldiers and explorers who descended on America shortly after the arrival of Christopher Columbus. In the New World, they battled the indigenous peoples for their gold and silver. People in parts of Mexico, Peru, and Bolivia were especially rich with these precious metals. Using more advanced weapons, including firearms, the conquistadores killed thousands of indigenous people. Many of those who weren't killed were forced to work in silver or gold mines. Meanwhile, the conquistadores shipped much of their gold back to the Spanish king, making Spain one of the wealthiest and most powerful countries in the world.

a lode. Gold veins form when hot fluids rise from within Earth and dissolve gold lodged in rock. The fluids find cracks in the rock and begin to cool. As a result, the gold is released out of the solution, forming gold-rich veins. The gold lodged inside a vein can vary in size from small grains to flakes to larger chunks called nuggets. Often, the minerals quartz and pyrite are also found in or near the vein.

When a river or stream passes over a gold vein near the earth's surface, some of the gold is washed away. This process, called erosion, takes

Streaks of gold are visible in this rock sample from the Haveri mine in Finland. Rocks that contain metal, like this one, are called ore. Ore from the Haveri mine has produced about 9,000 pounds (4,100 kg) of gold. The mine also produces large amounts of copper as well as the minerals pyrite and quartz.

many years. The gold that is washed away is deposited downstream along with sand and gravel. Over time, the gold grains and flakes collect in the riverbed or streambed. These areas are known as placer deposits.

The largest gold deposit ever discovered is a placer deposit in South Africa. Known as the Witwatersrand, it is the source of an estimated 40 percent of the world's gold. Large deposits of gold have been discovered in other countries as well, including Australia, Brazil, Canada, Russia, and the United States.

Panning is the least expensive way for a prospector to find gold. However, it is also tedious and physically demanding work that requires a lot of practice to master. The panning technique rarely leads to big finds. Today, most people who pan for gold do it for fun rather than profit.

Panning for Gold

The first prospectors found their gold in placer deposits. Working in rivers or streams, they used pans to scoop up water, sand, and gravel. When they swirled the contents of their pans, the less dense sand and gravel spilled out with the water. Since gold is more dense, it did not get swept up in the swirling water and remained at the bottom of the pan.

Panning can be done by a single prospector working alone. In addition, it does not require any fancy equipment. Because of this, panning has been popular throughout history, including during the famous California gold rush of the mid-nineteenth century.

Mining Gold

Removing gold from rocks is a much more difficult process. It requires many people and much more sophisticated equipment than panning. First, the gold-bearing rock, called ore, is dug out of the earth—a difficult task in itself. Then, the gold has to be extracted from the ore. A popular method of doing this is called cyanidation.

In the first step of cyanidation, machinery is used to crush the ore. Then a compound called cyanide is added, along with oxygen and water. Made out of carbon and nitrogen, the cyanide combines with the gold to create an ion that dissolves in the water. The rest of the ore, called gangue, does not dissolve in the water and can be discarded.

But now you have gold bound to cyanide and dissolved in water, which isn't very useful. How can the gold be recovered? The answer is by adding to the solution a metal that the cyanide bonds with more readily than gold. For example, when zinc is added to the water, the cyanide lets

Gold ore is made up of gold along with minerals and other metals that are less valuable. In order to refine gold ore, commercial mining companies use large industrial rock crushers. The gold processing plant seen above is located in the African country of Zimbabwe. Processing plants like this one are often located near gold mines in order to reduce the distance the ore has to travel.

go of the gold and attaches to the zinc. Gold is then precipitated out of the solution, or released, and collects on the bottom of the container. The water is poured off, leaving behind grains of pure gold.

Cyanidation may be effective at removing gold from ore, but it is also destructive to the environment. Cyanide is a dangerous chemical. Even in small doses, it can be fatal if ingested by humans. When used in gold mining, it can contaminate streams, rivers, and lakes in the area. Wildlife is endangered by the toxic cyanide in the wastewater. Humans can suffer, too, if the water supply becomes contaminated. Many environmental groups are calling for a ban on cyanide use in gold mining. Some states in the United States have already banned its use, as has the country of Turkey.

Bacteria and Gold

Cyanidation is not the only way miners reclaim gold from rock. Another method uses rock-eating bacteria. These tiny critters feed on pyrite, a type of rock often containing small amounts of gold. The bacteria eat the pyrite because it is made primarily of sulfur, one of their favorite foods. However, the bacteria cannot eat the gold and leave it untouched. Think about it as if someone offered you a fabulously delicious piece of chocolate pie. However, inside the pie were chunks of gold. Naturally, you would eat up all the pie but leave the chunks of gold on your plate. That is basically what the rock-eating bacteria do.

The technology of mining gold with bacteria has been around for more than twenty-five years, but its popularity has increased recently. For this technique, the bacteria, gold ore, water, and nutrients for the bacteria are mixed in a huge tank in an industrial plant. After the bacteria consume the sulfur in the tank, the solution still needs to be treated with additional chemicals, such as cyanide.

Alloys

Mixtures of two or more metals are called alloys. In nature, gold is often found in alloys with other metals, such as silver. Because the properties of gold are different from those of other metals, it is relatively easy to separate the metals in a gold alloy.

An alloy is a mixture and not a compound. This means that the elements in the mixture do not combine in fixed ratios. In a compound like water, hydrogen and oxygen atoms always combine in a ratio of two to one. In an alloy, however, atoms can combine in nearly any ratio. It is more like a mixture of two or more different colors of paint, or a mixture of milk and chocolate syrup to make chocolate milk.

Other Sources of Gold

Besides from within Earth and at the bottom of its streams and rivers, there are other places gold can be found. Scientists and prospectors are currently working on ways to make it profitable to extract gold from these sources.

Gold from Plants

Metals exist in small amounts in soil. When plants take up nutrients and water from the soil, they also take up tiny amounts of metal, including gold. Scientists have discovered plants that take up and store more metal than the average plant. These special plants, called hyperaccumulators, include leaf mustard, eucalyptus, canola, and alfalfa. Someday, these may be used as gold-gathering crops. Using plants to collect metals—a technique called phytomining—is still experimental. More research must be done to determine if it makes economic sense, but phytomining is

Fritz Haber *(left)* spent six years trying to recover gold from the oceans. Although these experiments failed, Haber is remembered as one of the most accomplished scientists of his time. In 1918, he won the Nobel Prize, the most prestigious award in science.

the subject of increasing interest around the world.

The practice of phytomining varies depending on the conditions. Usually, the plants are grown in an area near old mine tailings that still contain small amounts of gold. The roots of the plants take up the gold into their tissues as they draw nutrients. After the plants have stored as much gold as they can, they are harvested and burned. Gold is then recovered from the ashes.

An Ocean of Gold

It has been known for many years that the world's oceans contain more than a trillion dollars worth of gold. This fact inspired German chemist Fritz Haber (1868–1934) to attempt to mine the oceans for gold. He hoped that his profits would help his country pay off its debts from World War I (1914–1918). Unfortunately for him, Haber vastly overestimated the concentration of gold in seawater. In the end, the cost proved greater than the value of the gold he could extract.

In the future, perhaps a brilliant scientist or engineer will discover a better method of profitably gathering gold from the sea. Until then, ten million tons of gold are hidden underneath the surface, just out of reach of the eager prospector.

Chapter Five
A Useful Element

Gold's properties make it one of the most useful of all the metals. In this chapter, we'll look at a few of the many uses of gold in the arts, medicine, and industry.

Jewelry

Gold's ability to be bent and shaped into intricate designs makes it a favorite of jewelers and people who wear fine jewelry. Also, gold does not react with air or water, which makes it even more valuable. Unlike gold, other metals react with the oxygen in the air in a process known as oxidation. The rust that you can see on bridges and old cars is the result of the oxidation of iron. Copper is another metal that reacts with oxygen. If you have ever seen the Statue of Liberty, then you know what copper looks like when it oxidizes. Back in 1886, when the statue was unveiled, it was a reddish color. Over time, however, the surface of the metal reacted with oxygen and other elements in the atmosphere, forming a patina, the pale green film that you see today.

Had the Statue of Liberty been made out of gold, it would still be as shiny and golden today as the day it was created. In fact, in 1986 the city of New York decided to redo the statue's torch and cover it in gold. The

In 1916, the Statue of Liberty's torch was rebuilt so that it could be lit from inside, similar to a lighthouse. However, the new design eventually weakened the structure. In 1986, the torch was returned to its original design, and the flame was resurfaced with a thin layer of 24-karat gold. It now reflects sunlight from outside, as seen here.

golden torch will shine brightly and never dull as long as the statue exists.

The same properties that make gold an ideal metal for jewelry also make it good for coins. Gold, copper, and silver are the most common coinage metals. Today, gold coins are not used as money but instead are bought by collectors. The value of a gold coin fluctuates over time depending on the market price of gold and the demand for the coin among other collectors.

The Karat

Most gold used to make jewelry and coins is not 100 percent pure. That is because pure gold is too soft for these purposes. If pure gold were shaped into a ring, it would twist out of shape too easily. If it were made into a necklace, it would break without much difficulty. To solve this problem, jewelers usually use an alloy made of gold and another metal. Often, the other metal is silver or copper.

To measure the purity of gold, jewelers use karats. Twenty-four-karat gold is pure gold. Eighteen-karat gold is 75 percent pure, while 12-karat gold is 50 percent pure. In other words, to determine the percentage of pure gold in a sample, divide the number of karats by 24 and then multiply by 100.

Liquid Gold

Liquid gold is regularly used to decorate pottery and glass. It is prepared by mixing grains of gold with natural oils and other ingredients to create a thin paste of 4 to 12 percent gold. The liquid gold is then painted onto the surface of the object. The object is heated to burn off the oil and other ingredients, leaving behind a 22-karat gold finish.

Plates, cups, and perfume bottles are just a few of the products that can be enhanced with liquid gold decorations.

Gold Leaf

Gold's ability to be pounded into thin sheets allows a little bit to go a long way. When gold is worked into a very thin foil, it is called gold leaf. If an artist wants to add a gold finish to a sculpture, or if an architect wants to add gold to a fixture on the outside of

Throughout history, artisans around the world have used gold in their designs. In a monastery in Myanmar, Southeast Asia, a gold-covered Buddha statue *(right)* is a shining example of how gold can be used to create beautiful, yet durable, art objects.

a building or home, then gold leaf is the material of choice. Instead of making a piece entirely of gold, a small amount of gold can be pounded into gold leaf and wrapped tightly around the piece. Although the piece might look like it is made out of gold, it really has only a thin gold wrapping. Gold leaf gives the appearance of solid gold without the hefty price tag.

Gold in Medicine

Gold forms a compound named gold sodium thiomalate, also known as Myocrisin. This compound, made of gold, sodium, oxygen, and sulfur, was initially used as a drug to treat tuberculosis. Although it wasn't very effective in treating this disease, doctors accidentally discovered another use for the drug. Myocrisin, they found, helped reduce the severity of symptoms of arthritis, a disease in which the joints in the legs and hands swell and cause pain. Although no one really knows for sure why it works, Myocrisin and a similar gold compound known as auranofin have been used to treat arthritis since the 1960s.

Gold in Your Mouth

Dentists use more than 60 tons (54 metric tons) of gold every year. Gold used in dentistry is always alloyed with other metals. Often these metals are silver, palladium (Pd), copper, and zinc. The mixture chosen has to be durable and easy to work with. Gold is the main ingredient because it will not react with any bodily fluid or anything put into the mouth. Also, if it were accidentally swallowed, the patient wouldn't be harmed. That's because gold is completely nontoxic inside the body.

Gold is especially popular for more complex dental work such as inlays, crowns, and bridges. These require an alloy that is strong but malleable enough to be molded specifically for the patient. A gold alloy is the perfect choice for such a task.

A dental technician uses a grinding wheel to shape a gold tooth. A cast of the patient's teeth rests to his right. Gold is often used to make a crown, a tooth-shaped cap that covers a broken or decayed tooth. Gold crowns are strong enough to withstand years of biting and chewing. They need to be ground smooth so that food and bacteria do not stick to the surface of the crown.

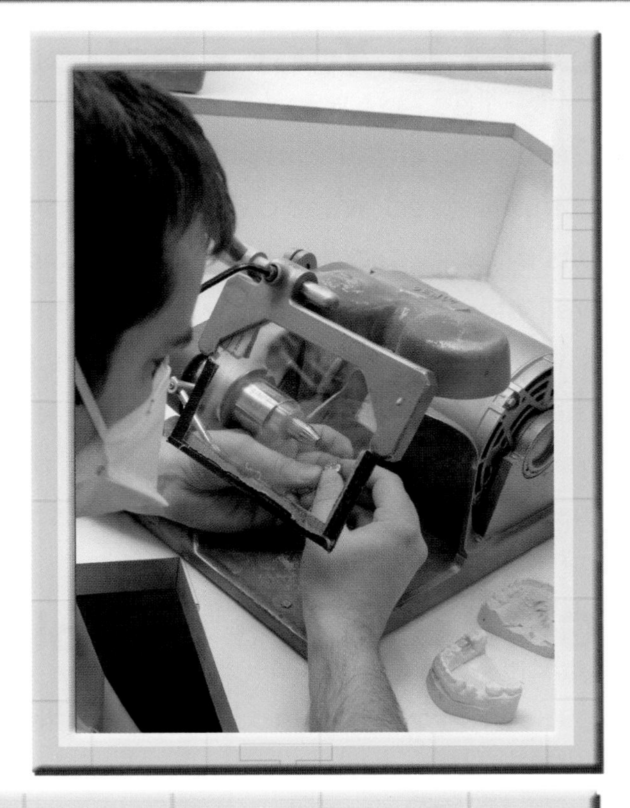

Gold Compounds

As mentioned earlier, gold does not easily combine with other elements to form compounds. This is because its outer electron is held tight by the nucleus of the atom and not made available for bonding. Despite its tight grip on its outer electron, gold can indeed form compounds. One of these compounds, Myocrisin, has already been described.

Two other gold compounds are auric chloride and chlorauric acid. ("Auric" is another way of saying "gold.") Auric chloride ($AuCl_3$) is made of gold and chlorine. It is used in the synthesis of other gold compounds, and it is also used to start some reactions in organic chemistry. Chlorauric acid ($HAuCl_4$) consists of gold, hydrogen, and chlorine. Chlorauric acid, which is more common than auric chloride, is used in photography during the developing process.

Gold in Your Electronics

Electronic circuits in telephones, televisions, computers, automobiles, computerized wheelchairs, and communications satellites often use tiny bits of gold. Gold is a popular choice because it conducts electricity very well at very high and low temperatures. It also is resistant to corrosion, so it will last a long time. Sometimes, thin layers of gold are placed over cheaper metals (like copper) to protect them from being damaged by oxygen and water. In any electronic device where performance is critical, there is a good chance that gold is used in its circuitry.

Gold to Keep It Cool

The sun's rays are made out of different types of radiation. Some of it is white light, which can be broken apart into the colors of the visible light spectrum. The sun also emits infrared radiation, or heat. Gold has the ability to reflect this type of radiation. Some architects take advantage of this property of gold to build energy-efficient buildings. Their designs use windows that have a thin layer of gold in the glass. Visible light passes through, but the gold in the glass reflects the infrared radiation and keeps the room from heating up. This might seem like an expensive solution, but if the building is big enough, it can save a lot of money in air-conditioning costs.

The Future of Gold

Gold's properties make it a very popular element. It is a favorite of jewelry makers, artists, chemists, dentists, manufacturers, and engineers. All of these professions have found ways to make gold useful.

The need for gold is likely to increase as people discover new uses for it. However, there is a limited supply of gold in rocks, rivers, and streams. Eventually, these sources could be depleted of nearly all their gold.

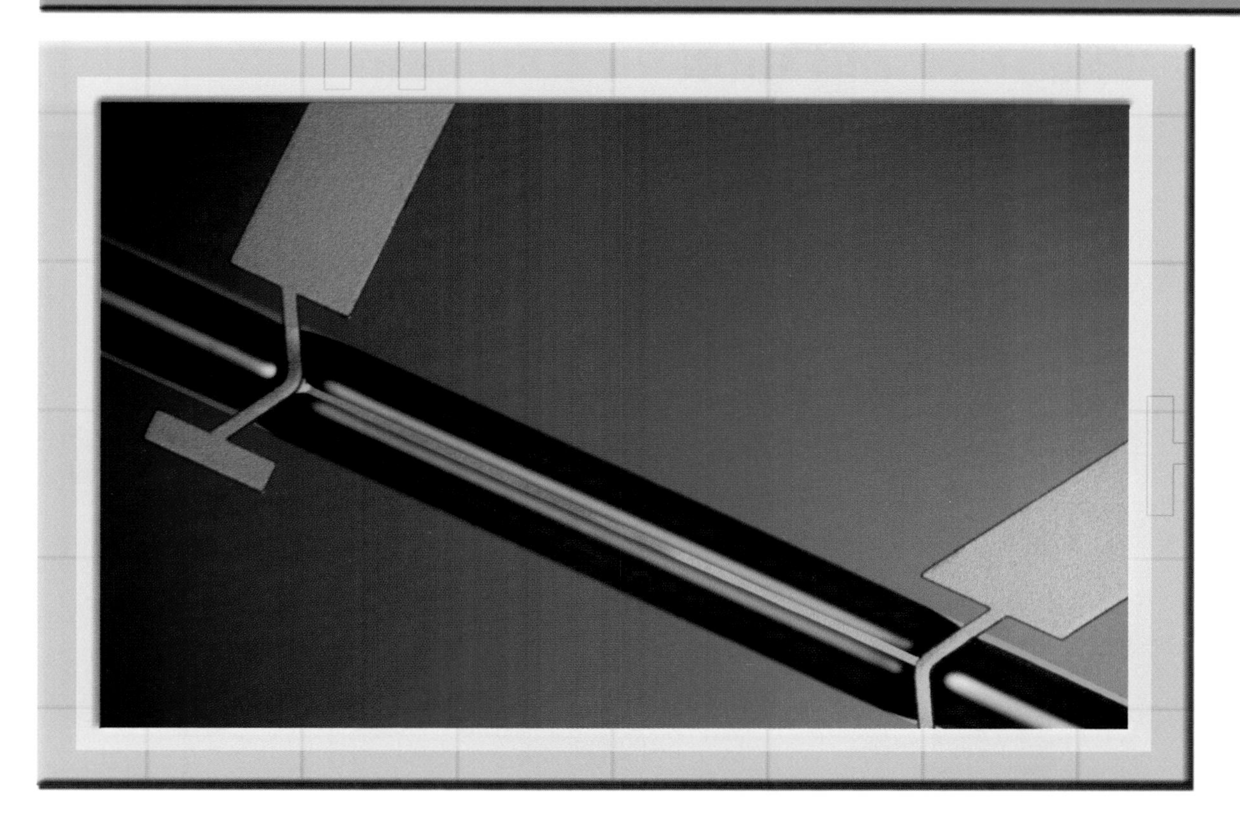

A thermal conductivity detector *(above)* contains a gold wire thinner than a human hair. In this photograph, the gold wire *(upper left to lower right)* has been magnified to 850 times its actual size. The detector is used to measure how well gases conduct heat. The technology used to create these miniature detectors is expected to become increasingly popular.

As gold becomes less available, gold recycling will become more important. Computers, telephones, and other electronic devices can be stripped of their gold when they are no longer useful. Even though each of these devices has only a tiny bit of gold inside, the gold can add up when thousands of these devices are recycled.

Its many uses make gold an element with a bright future. Recycling will make its future even brighter.

The Periodic Table of Elements

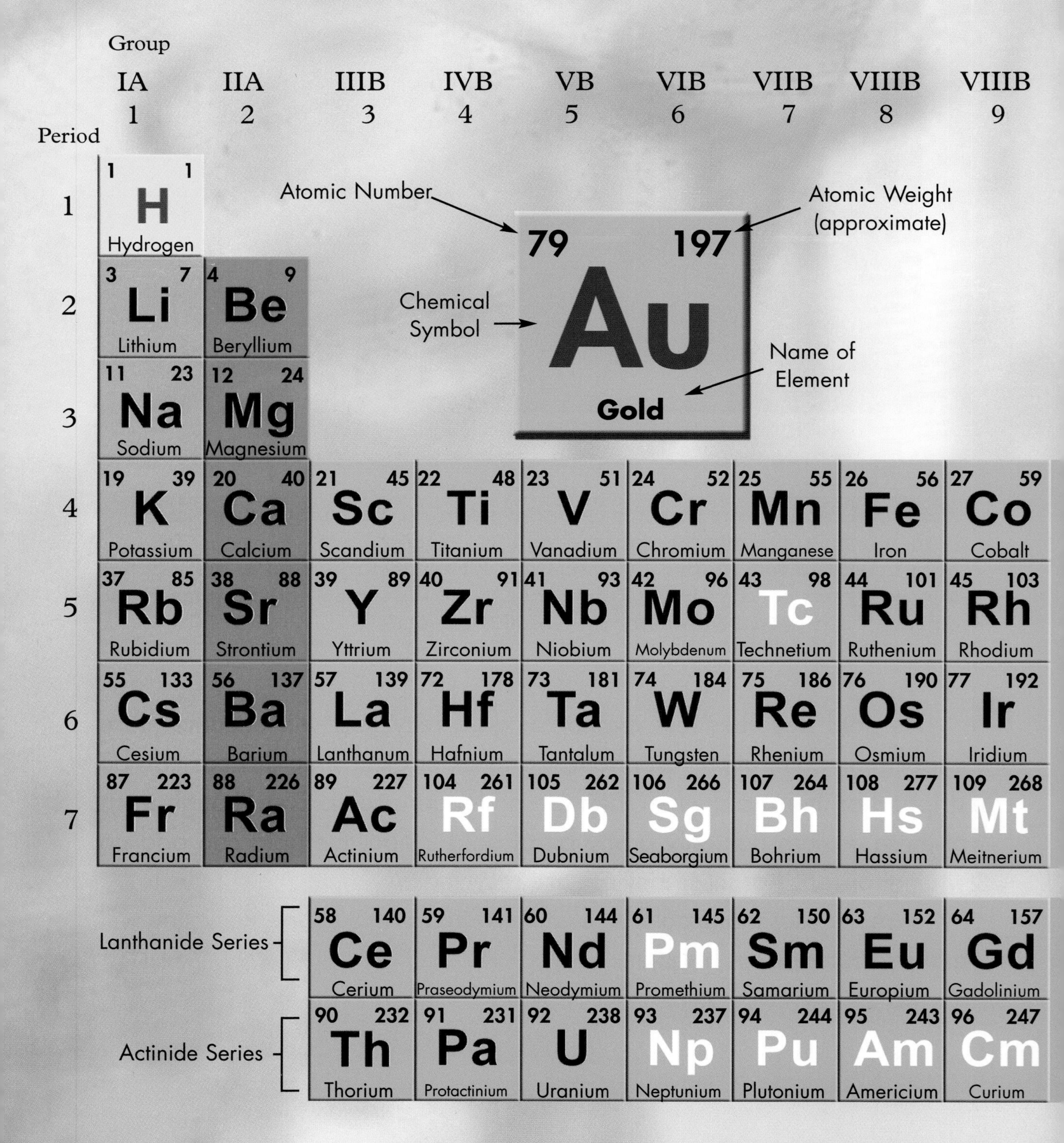

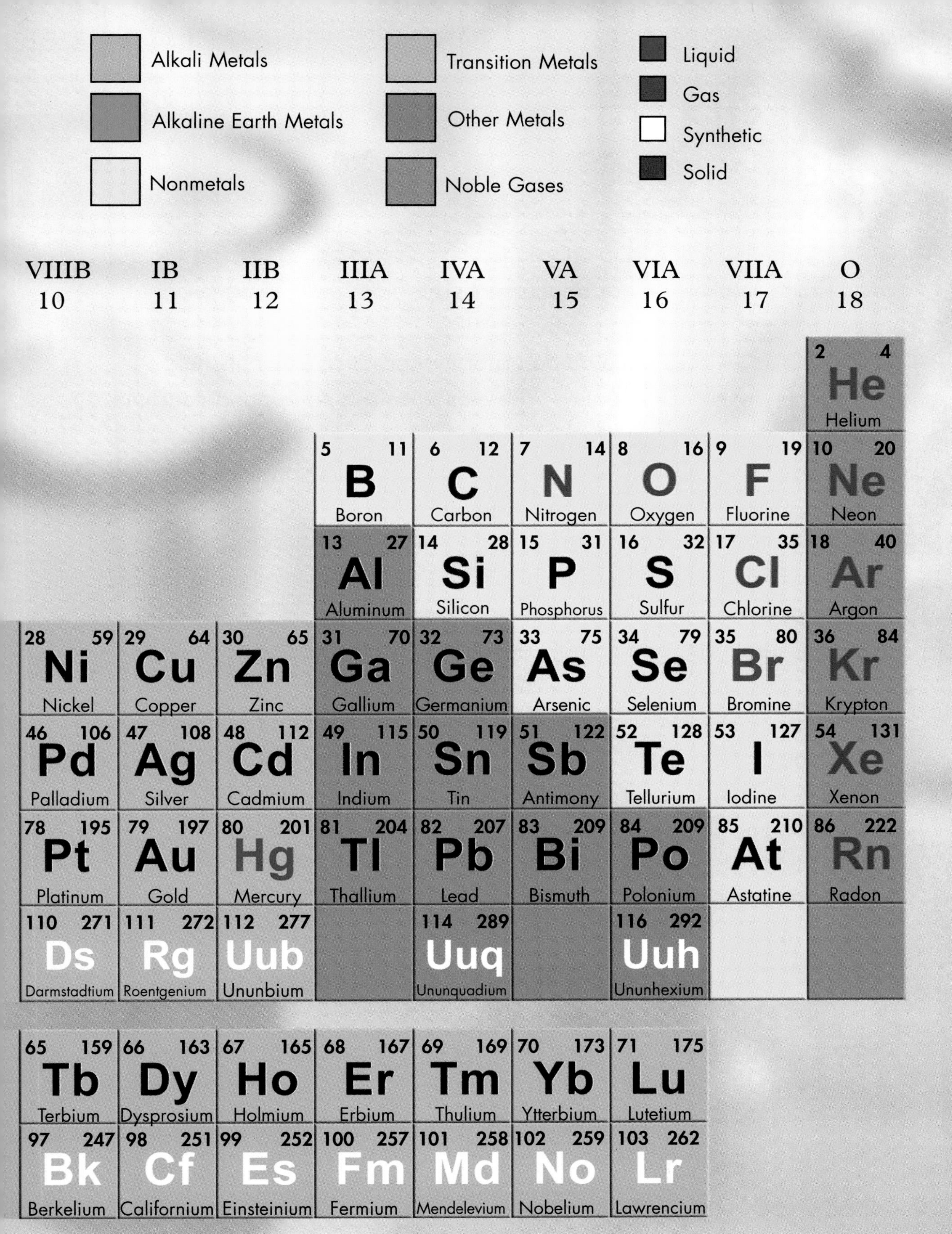

Alkali Metals
Alkaline Earth Metals
Nonmetals
Transition Metals
Other Metals
Noble Gases
Liquid
Gas
Synthetic
Solid
VIIIB 10
IB 11
IIB 12
IIIA 13
IVA 14
VA 15
VIA 16
VIIA 17
O 18
2 4 He Helium
5 11 B Boron
6 12 C Carbon
7 14 N Nitrogen
8 16 O Oxygen
9 19 F Fluorine
10 20 Ne Neon
13 27 Al Aluminum
14 28 Si Silicon
15 31 P Phosphorus
16 32 S Sulfur
17 35 Cl Chlorine
18 40 Ar Argon
28 59 Ni Nickel
29 64 Cu Copper
30 65 Zn Zinc
31 70 Ga Gallium
32 73 Ge Germanium
33 75 As Arsenic
34 79 Se Selenium
35 80 Br Bromine
36 84 Kr Krypton
46 106 Pd Palladium
47 108 Ag Silver
48 112 Cd Cadmium
49 115 In Indium
50 119 Sn Tin
51 122 Sb Antimony
52 128 Te Tellurium
53 127 I Iodine
54 131 Xe Xenon
78 195 Pt Platinum
79 197 Au Gold
80 201 Hg Mercury
81 204 Tl Thallium
82 207 Pb Lead
83 209 Bi Bismuth
84 209 Po Polonium
85 210 At Astatine
86 222 Rn Radon
110 271 Ds Darmstadtium
111 272 Rg Roentgenium
112 277 Uub Ununbium
114 289 Uuq Ununquadium
116 292 Uuh Ununhexium
65 159 Tb Terbium
66 163 Dy Dysprosium
67 165 Ho Holmium
68 167 Er Erbium
69 169 Tm Thulium
70 173 Yb Ytterbium
71 175 Lu Lutetium
97 247 Bk Berkelium
98 251 Cf Californium
99 252 Es Einsteinium
100 257 Fm Fermium
101 258 Md Mendelevium
102 259 No Nobelium
103 262 Lr Lawrencium

Glossary

alloy A mixture of two or more metals.

atom The smallest unit of an element. The atom itself is made out of electrons, protons, and neutrons.

compound A substance made up of two or more elements held together by chemical bonds. The elements in a compound combine in fixed ratios, such as two atoms to one, or three to two, etc.

conduct To allow something to pass through; gold conducts both heat and electricity.

corrosion A chemical reaction that results in the deterioration of a metal. Often, metals corrode when they react with oxygen in the air, a process known as oxidation.

density The mass of a sample divided by its volume.

ductile Capable of being stretched into a wire.

electron A negatively charged particle found outside of the nucleus, or center, of an atom.

ion A positively or negatively charged atom.

karat A measure of the purity of gold. Pure gold is 24 karats.

malleable Capable of being bent or shaped.

neutron A particle without charge that is part of the nucleus of most atoms.

ore Rock that contains a valuable metal or mineral.

prospector One who explores for valuable metals, minerals, or oil.

proton A positively charged particle that is part of the nucleus of an atom.

radiation Energy that moves in the form of waves, rays, or particles.

tuberculosis An infectious disease that affects the lungs and results in coughing, fever, fatigue, and weight loss.

For More Information

Center for Science and Engineering Education
Lawrence Berkeley National Laboratory
1 Cyclotron Road MS 7R0222
Berkeley, CA 94720
(510) 486-5511
Web site: http://www.lbl.gov

International Union of Pure and Applied Chemistry
IUPAC Secretariat
P.O. Box 13757
Research Triangle Park, NC 27709-3757
(919) 485-8700
Web site: http://www.iupac.org

Los Alamos National Laboratory
P.O. Box 1663
Los Alamos, NM 87545
(888) 841-8256
Web site: http://periodic.lanl.gov

Web Sites

Due to the changing nature of Internet links, the Rosen Publishing Group, Inc., has developed an online list of Web sites related to the subject of this book. This site is updated regularly. Please use this link to access the list:

http://www.rosenlinks.com/uept/gold

For Further Reading

Angliss, Sarah. *Gold.* Tarrytown, NY: Benchmark, 2000.

Baldwin, Carol. *Metals.* Chicago, IL: Raintree, 2004.

Knapp, Brian. *Copper, Silver, and Gold.* Danbury, CT: Grolier, 2002.

Oxlade, Chris. *Metals.* Chicago, IL: Heinemann, 2002.

Stwertka, Albert. *A Guide to the Elements.* New York, NY: Oxford University Press, 2002.

Bibliography

Acevedo, Fernando. "Present and Future of Bioleaching in Developing Countries." *Electronic Journal of Biotechnology*, Vol. 5, No. 2, August 15, 2002.

Atkins, P. W. *The Periodic Kingdom: A Journey into the Land of the Chemical Elements.* New York, NY: Basic Books, 1995.

Ball, Philip. *The Ingredients: A Guided Tour of the Elements.* Oxford, England: Oxford University Press, 2002.

Barrett, Jack. *Atomic Structure and Periodicity.* Hoboken, NJ: John Wiley & Sons, 2002.

Emsley, John. *Nature's Building Blocks: An A–Z Guide to the Elements.* Oxford, England: Oxford University Press, 2001.

Gold Institute. "Nothing Works Like Gold." Retrieved August 3, 2005 (http://www.goldinstitute.org).

Pellant, Chris. *Rocks and Minerals.* New York, NY: Dorling Kindersley, 1992.

World Gold Council. "Industrial Applications." Retrieved August 3, 2005 (http://www.gold.org/discover/sci_indu/indust_app/index.html).

Zumdahl, Steven S. *Chemistry.* Lexington, MA: D.C. Heath and Company, 1989.

Index

About the Author

Brian Belval has a bachelor's degree in biochemistry from the University of Illinois. He worked as a research scientist for a number of years before returning to school to study literature at the University of Massachusetts. He currently lives in New York City, where he combines his interest in science and writing as an editor of young adult nonfiction.

Photo Credits

Cover, pp. 1, 8, 14, 16, 42–43 by Tahara Anderson; p. 5 © Jacana/Photo Researchers, Inc.; p. 7 © Reid Goldsborough; p. 13 © Novosti/Photo Researchers, Inc.; p. 19 © Charles D. Winters/Photo Researchers, Inc.; p. 21 © AP/Wide World Photos; p. 23 © Maximilian Stock, Ltd./Photo Researchers, Inc.; pp. 24, 26 by Maura McConnell; p. 29 by Jari Vaatainen, Geological Survey of Finland; p. 30 © Phil Schermeister/Corbis; p. 31 © Peter Bowater/Photo Researchers, Inc.; p. 34 © Topical Press Agency/Getty Images; p. 36 © Ron Watts/Corbis; p. 37 © Sergio Pitamitz/zefa/Corbis; p. 39 © John McLean/Photo Researchers, Inc.; p. 41 © Volker Steger/Photo Researchers, Inc.

Special thanks to Megan Roberts, director of science, Region 9 Schools, New York City, NY, and Jenny Ingber, high school chemistry teacher, Region 9 Schools, New York City, NY, for their assistance in executing the science experiments illustrated in this book.

Designer: Tahara Anderson; Editor: Christopher Roberts